E 332.4 MOL
Molter, Carey, 1973-
How much is $50.00?
Carver County Library

How Much Is $50.00?

Carey Molter

Consulting Editor, Monica Marx, M.A./Reading Specialist

Published by SandCastle™, an imprint of ABDO Publishing Company, 4940 Viking Drive, Edina, Minnesota 55435.

Copyright © 2003 by Abdo Consulting Group, Inc. International copyrights reserved in all countries. No part of this book may be reproduced in any form without written permission from the publisher. SandCastle™ is a trademark and logo of ABDO Publishing Company.
Printed in the United States.

Credits
Edited by: Pam Price
Curriculum Coordinator: Nancy Tuminelly
Cover and Interior Design and Production: Mighty Media
Photo Credits: Hemera Studio, PhotoDisc, Rubberball Productions, Stockbyte

Library of Congress Cataloging-in-Publication Data

Molter, Carey, 1973-
    How much is $50.00? / Carey Molter.
      p. cm. -- (Dollars & cents)
    Includes index.
    Summary: Explains what a fifty dollar bill is, how it compares to other dollar bills, and how many fifties are needed to purchase different items.
    ISBN 1-57765-890-6
    1. Money--Juvenile literature. 2. Dollar, American--Juvenile literature. 3. Addition--Juvenile literature. [1. Money.] I. Title: How much is fifty dollars?. II. Title. III. Series.

HG221.5 .M6545 2002   2002071708
332.4'973--dc21

SandCastle™ books are created by a professional team of educators, reading specialists, and content developers around five essential components that include phonemic awareness, phonics, vocabulary, text comprehension, and fluency. All books are written, reviewed, and leveled for guided reading, early intervention reading, and Accelerated Reader® programs and designed for use in shared, guided, and independent reading and writing activities to support a balanced approach to literacy instruction.

Let Us Know

After reading the book, SandCastle would like you to tell us your stories about reading. What is your favorite page? Was there something hard that you needed help with? Share the ups and downs of learning to read. We want to hear from you! To get posted on the ABDO Publishing Company Web site, send us email at:

sandcastle@abdopub.com

SandCastle Level: Beginning

This is a fifty-dollar bill.

Fifty dollars is the same as one hundred half-dollars.

This is how to write fifty dollars.

$50.00

Fifty dollars is the same as 50 one-dollar bills.

Fifty dollars is the same as ten five-dollar bills.

Fifty dollars is the same as two twenty-dollar bills plus one ten-dollar bill.

This ball costs $50.00.

That is one fifty-dollar bill.

This lamp costs $100.00.

That is two fifty-dollar bills.

This guitar costs $150.00.

That is three fifty-dollar bills.

How many dollars does one balloon ride cost?

(two hundred)

Picture Index

ball, p. 15

fifty dollars,
pp. 3, 7, 9, 11, 13

fifty-dollar bill, pp. 5, 15

lamp, p. 17

More about the Fifty-Dollar Bill

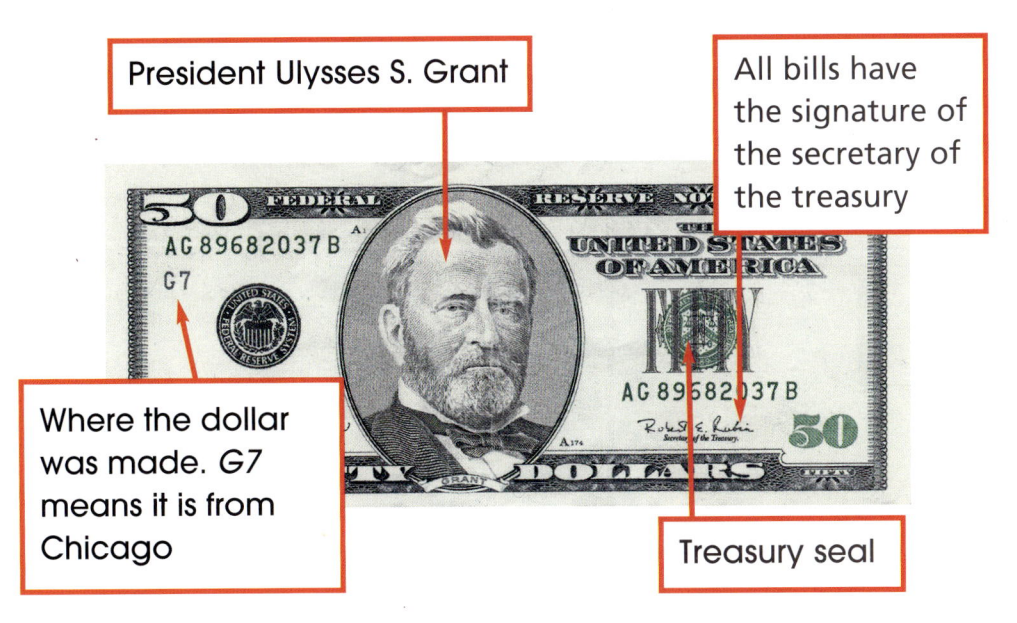

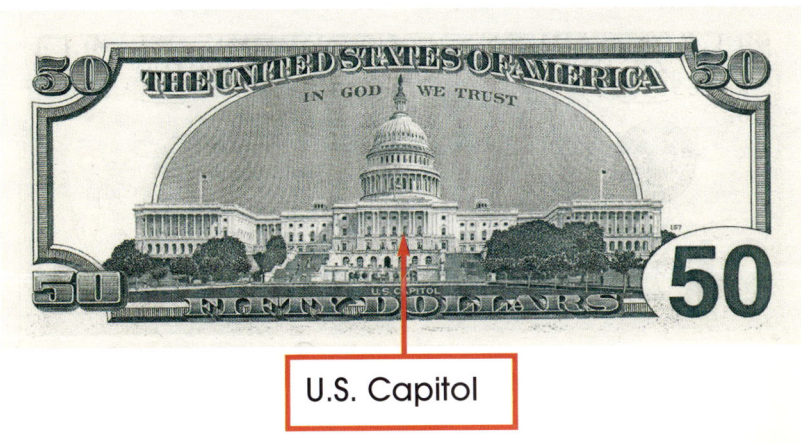

President Ulysses S. Grant

All bills have the signature of the secretary of the treasury

Where the dollar was made. *G7* means it is from Chicago

Treasury seal

U.S. Capitol

23

About SandCastle™

A professional team of educators, reading specialists, and content developers created the SandCastle™ series to support young readers as they develop reading skills and strategies and increase their general knowledge. The SandCastle™ series has four levels that correspond to early literacy development in young children. The levels are provided to help teachers and parents select the appropriate books for young readers.

Emerging Readers
(no flags)

Beginning Readers
(1 flag)

Transitional Readers
(2 flags)

Fluent Readers
(3 flags)

These levels are meant only as a guide. All levels are subject to change.

To see a complete list of SandCastle™ books and other nonfiction titles from ABDO Publishing Company, visit **www.abdopub.com** or contact us at:

4940 Viking Drive, Edina, Minnesota 55435 • 1-800-800-1312 • fax: 1-952-831-1632

JAN 2 4 2006